Generis

PUBLISHING

REGULATED ACCUMULATION
Spectral Geometry, Physical Structure, and Stability across Scales

Savio Antonio Vogt

Title: **REGULATED ACCUMULATION**
Spectral Geometry, Physical Structure, and Stability across Scales

ISBN: 979-8-89966-428-1

Author: Savio Antonio Vogt

Cover image: AI Generated

Publisher: Generis Publishing
Online orders: www.generis-publishing.com
Contact email: info@generis-publishing.com

REGULATED ACCUMULATION

Spectral Geometry, Physical Structure, and Stability across Scales

Mr. Savio Antonio Vogt

CONTENTS

Structural Integrity of the Monograph

The architecture now reads as,

Foundational Diagnosis → Physical Consequence → Empirical Consolidation → Equation Unification → Philosophical Closure

The hierarchy is preserved,

Geometry → Accumulation → Density → Operator → Physics → Interpretation

No inversion.

TERMINOLOGY/DICTIONARY

I. Foundational Structural Terms

These are the primitive commitments of the monograph,

- Geometric Saturation.
- Accumulation Principle.
- Unbounded Additive Continuation.
- Structural Compatibility.
- Legitimacy Condition.
- Admissible Extension.
- Spectral Substrate.
- Density Thinning.
- Logarithmic Compression.
- Upper-Bound Regulation.
- Pre-collapse Structure.

These form the metaphysical–mathematical backbone of the work.

II. Regulator Vocabulary

Terms directly tied to the central regulating mechanism,

- The Regulator.
- Saturation before divergence.
- Bounded deviation.
- Spectral tapering.
- Geometric moderation.
- Operator admissibility.
- Compact resolvent behavior.
- Infrared saturation.
- Bounded transfer.
- Spectral compression.
- Constrained accumulation.

III. Spectral & Analytical Framework

Recurring mathematical language across domains,

- Spectral Density.

- Operator Theory.
- Resolvent Structure.
- Energy Cascade
- Confinement
- Singularity Formation.
- Arithmetic Density.
- Scale Resemblance.
- Cross-domain Scaling.
- Spectral Spacetime.
- Substrate Regulation.

IV. Methodological Terms (Chapter 17 emphasis)

These describe the philosophical stance,

- Infinite continuation.
- Ad hoc renormalization.
- Artificial infinities.
- Improvised truncation.
- Perturbative control.
- Local precision.
- Numerical simulation.
- Extension admissibility.
- Structural discipline.
- Methodological restraint.

V. Cross-Domain Unification Terms (Chapter 16)

Used in the Structural Unity table,

- Prime Density – Log thinning.
- Yang–Mills – Confinement.
- Navier–Stokes – Energy cascade.
- General Relativity – Spectral pre-collapse.
- Operator compactness – Density thinning.

These terms anchor the claim that the regulator is not domain-specific but structurally emergent.

VI. Clarificatory / Defensive Terms (Ch. 14–15)

To prevent misinterpretation,

- Dismissed anomalies.
- Recurrent bounded deviations.
- Structural vs. empirical proof.
- Interpretative lens.
- Not metaphysical cosmology.
- Not alternative physics.
- Not imposed cutoff.
- Not numerical trick.

Condensed Core Vocabulary (Most Central Terms)

If reduced to the 15 most essential terms, the monograph revolves around,

1. Geometric Saturation.
2. Accumulation Principle.
3. Structural Compatibility.
4. Legitimacy Condition.
5. Spectral Regulation.
6. Logarithmic Compression.
7. Density Thinning.
8. Bounded Transfer.
9. Infrared Saturation.
10. Operator Admissibility.
11. Pre-collapse Structure.
12. Constrained Continuation.
13. Structural Unity.
14. Methodological Restraint.
15. Admissible Extension.

PREFACE

Modern theoretical physics and mathematics are extraordinarily successful. Yet across multiple domains, quantum field theory, fluid mechanics, spectral theory, cosmology, one encounters a recurring structural difficulty: uncontrolled accumulation.

- Divergences appear.
- Renormalization becomes necessary.
- Cut-offs are introduced.
- Regularizations are imposed.
- Statistical corrections are applied.

These techniques work are powerful. But structurally they often operate after divergence has already manifested. The work collected in the accompanying series of preprints proposes a different viewpoint,

That accumulation itself must be geometrically regulated at the structural level.

The programme does not reject existing physics. It does not replace established mathematics. It does not modify successful equations. Instead, it proposes that many divergences arise because additive growth is applied where multiplicative saturation is geometrically required.

Across thirty-two preprints, this idea has been developed through,

- Operator theory.
- Spectral geometry.
- Functional analysis.
- Quantum field structure.
- Navier–Stokes energy cascades.
- Yang–Mills confinement.
- Cosmological scaling.
- Statistical consistency testing.

The equations remain intact, only the structure around them is refined. This monograph is not a repetition of those papers. It is an explanatory synthesis. Its aim is clarity. To answer a simple question,

Where does this work belong?

The answer, argued here is,

- Theoretical Physics.
- Mathematical Physics.
- Pure Mathematics.
- Applied Mathematics.
- Statistical Validation.

It is a structural programme.

This text presents the framework in a unified philosophical and geometric order, accessible to specialists and interested readers alike.

- No equation is altered.
- No claim is exaggerated.
- Every concept is situated within existing mathematical language.

The reader is invited not to accept conclusions, but to examine structure.

DEDICATION & ACKNOWLEDGEMENTS

Dedication

First and foremost, I dedicate this work to **God Almighty**, whose sustaining grace granted me strength through more than a decade marked by rejection, uncertainty, and perseverance. Without that quiet constancy, this monograph would not have come into being.

To my beloved parents, **Lester Marshall Bernard Vogt** and **Edna Antoinette Vogt (née Vaz)** your endurance, sacrifice, and faith remain the unseen architecture beneath all that I undertake. Despite the trials our family has endured, I trust that this work, by God's grace, may speak meaningfully to hearts and minds beyond our immediate horizon.

Acknowledgements

I record my deep gratitude to my alma mater, **Loyola College, Chennai**, whose Jesuit tradition of holistic formation shaped both intellect and conscience in equal measure.

I am equally indebted to **The University of Leicester** for its rigorous research training and intellectual discipline. In particular, I acknowledge with sincere appreciation Professor Nigel P. Bannister and Professor Richard Ambrosi for their illuminating lectures and scholarly example.

I extend heartfelt thanks to the editors of **ScienceOpen** for providing an open platform through which my academic preprints could reach a global readership. Their commitment to accessible scholarship has been instrumental in the evolution of this work.

My appreciation also goes to the Editorial Team of **Generis Publishing** for recognizing the depth and contributory intention of this research and for inviting its publication in monograph form. I value the trust placed in this collaboration and look forward to its continued development.

To independent researchers everywhere: persevere. Scholarship is not confined to institutional circumstance, funding structure, or geography. Let your work reflect your life with integrity; what is authentically built cannot be taken away. Even beyond the familiar pressures of "publish or perish," meaningful contributions may emerge from the humblest origins. When one path closes, adapt, refine, and continue, there is dignity in the evolution of thought.

This monograph does not seek confrontation, but clarification. It stands in respect of established scholarship while quietly affirming that intellectual depth may arise wherever disciplined thought is sincerely pursued.

میری پیاری عشرت، آپ جہاں کہیں بھی ہوں، میں آپ کو جاننا چاہتا ہوں کہ میں آپ سے پیار کرتا ہوں اور یہ کام آپ کے لیے وقف کرتا ہوں۔

RESEARCH CLASSIFICATION

Primary Domain

Mathematical & Theoretical Physics

Secondary Domains

Pure & Applied Mathematics

- Spectral Geometry.
- Operator Theory.
- Quantum Field Theory.
- Fluid Mechanics.
- Cosmology.
- Applied Statistical Methods.

Methodology

Geometric structural reformulation with operator-theoretic validation and empirical comparison.

LIST OF FIGURES & TABLES

TABLE HEADINGS

FIGURE HEADINGS

PART I.
FOUNDATIONAL IDEA

CHAPTER 1
The Problem of Unregulated Accumulation

1.1 A Recurrent Structural Pattern

Across mathematics and physics, growth processes are typically treated additively.

- Energy accumulates.
- Modes sum.
- Spectra extend.
- Probability densities integrate.
- Curvature compounds.

Additivity is natural. It is intuitive and it aligns with linear superposition however, in many systems, additive continuation produces divergence,

- Ultraviolet divergence in quantum field theory.[12]
- Infrared divergence in gauge theories.[9][10]
- Blow-up scenarios in fluid equations.[7][19]
- Non-compact spectral growth.[4]
- Cosmological singular behavior.[11]

The response has historically been technical,

- Regularization.
- Renormalization.[12]
- Boundary conditions.
- Cut-offs.
- Distributional reinterpretation.

These are powerful tools and they remain valid, but they are reactive. The present framework asks a prior question,

Is unbounded additive accumulation structurally correct at all scales?

1.2 Additive Growth versus Geometric Saturation

Consider two structural modes of accumulation,

Additive,

$$S_n = \sum_{k=1}^{n} a_k$$

Multiplicative,

$$P_n = \prod_{k=1}^{n} (1 + b_k)$$

Additive growth is linear in structure and Multiplicative growth carries intrinsic curvature. When translated into spectral language, additive growth corresponds to unbounded eigenvalue summation consistent with classical asymptotics.[4] Multiplicative regulation introduces density compression.

The proposal developed in the preprints is,

Physical systems behave as if geometric saturation is embedded within accumulation.

Not imposed externally, but structurally present.

1.3 Divergence as Structural Mismatch

Many divergences can be interpreted as applying additive continuation beyond geometric validity.

Some examples are,

- Perturbative QFT expansions.[12]
- Navier–Stokes cascade extension.[7][8]
- Spectral trace divergence.[26]
- Mode density over-flow.

In each case, the mathematics is correct however, the structure is incomplete.[11]

The regulator introduced in the series, termed the Additive–Multiplicative (AM) Regulator, proposes a hybrid accumulation form preserving local linearity while enforcing global spectral compactness.[5][16]

The detailed equations appear in subsequent chapters.

1.4 Spectral Compactness as a Stability Principle

- Compact operators possess discrete spectra accumulating only at infinity.[5][17]
- Compact resolvents ensure controlled eigenvalue growth.[26]
- Self-adjointness enforces physical observability.[16]

These are not arbitrary mathematical properties. They encode stability.

The central thesis of this programme is,

Physical law prefers compact spectral behavior.

When equations appear divergent, the divergence signals missing geometric regulation, not failure of the law itself.

1.5 Direction of the Prog ramme

The research direction unfolds in four stages,

- Define regulated accumulation formally.
- Prove operator-theoretic consistency.[5][16][18]
- Embed the structure into physical equations without altering them.
- Compare predictions with empirical data across scales.[24][25]

The forthcoming chapters follow this order.

FOUNDATIONAL IDEA

CHAPTER 2

The Meaning of *Scale and Aggregation*

2.1 Scale Is Not Merely Size

In mathematics, scale is often treated parametrically. A variable increases. A domain extends. A spectrum enlarges. In physics, scale corresponds to energy, distance, curvature, or frequency. Yet scale is not simply magnitude. It is structural resolution.

A system may behave linearly at small scale and non-linearly at large scale. It may appear continuous locally and discrete globally. It may accumulate additively in one regime and multiplicatively in another. Classical turbulence scaling already reflects such regime dependence.[8]

The central claim developed across the preprints is that scale transition requires regulation. Unregulated continuation across scales produces artifacts that are then corrected externally.

2.2 Accumulation as a Structural Act

Accumulation is not neutral.

Whenever it is written,

$$\sum_{n=1}^{\infty} a_n$$

It is assumed, additive legitimacy at all orders.

And once again, when it is written,

$$\int_0^{\infty} f(x)\,dx$$

Normally assuming density remains integrable.

But physical systems do not guarantee infinite additive validity.

- In spectral systems, density increases with eigenvalue magnitude.[4]
- In turbulent systems, energy cascades downward.[8]
- In quantum fields, modes proliferate.[12]

The question becomes,

What constrains accumulation naturally?

The proposed answer is,

Geometric Saturation.

2.3 Hybrid Regulation

The AM-Regulator framework proposes a hybrid structure,

- Local additivity.
- Global multiplicative compression.

This preserves,

- Perturbative correctness.[12]
- Linearized physics.
- Known field equations.

While enforcing,

- Spectral compactness.[5]
- Density control.
- Stability under scaling.

The regulator is not imposed numerically. It is embedded structurally.

2.4 Philosophical Position

This programme does not deny renormalization. It suggests renormalization reflects deeper geometric regulation already implied by spectral compactness.[12][15] In this view, divergence is diagnostic. It signals a structural mismatch between additive modeling and geometric necessity.

FOUNDATIONAL IDEA
CHAPTER 3
The Additive–Multiplicative Regulator
Structural Governance of Accumulation

3.1 The Intellectual Misplacement of the Regulator

Across the preprint sequence, particularly,

- *Scale-Regulated Aggregation: A Structural Link between Prime Distributions, Fluid Dynamics, and Gravitation*
- *Foundations of the AM-Regulator: Deep Structure, Duality, Compactness, and Functorial Degeneracy*
- *Operator-Theoretic Foundations of Regulated Accumulation*

A recurring interpretive error has emerged.

Critics could possibly attempt to,

- Insert the regulator into existing additive formalism.
- Derive it using Euclidean asymptotics.
- Treat it as a modified renormalization constant.[12]
- Reduce it to a numerical damping function.

All of these approaches reverse the logical order. The regulator is not a correction within additivity. It is a condition governing when additivity is legitimate.

3.2 Accumulation Is Never Neutral

In classical mathematics,

$$S_n = \sum_{k=1}^{n} a_k$$

The notation hides an assumption that, additivity may continue indefinitely without structural change. Yet geometric systems do not behave that way.

Clear cut examples of this are,

- On a sphere, locally parallel geodesics converge.[4]
- In turbulent flow, energy cascades dissipate at finite scale.[8]
- In prime distribution, density thins logarithmically.[1][2]
- In quantum fields, mode density increases with energy.

Additivity is always embedded in geometry. The AM-Regulator makes this embedding explicit.

3.3 Formal Hybrid Structure

Across the mathematical formalism papers, regulated accumulation appears abstractly as,

$$R(n) = A(n)\Phi(M(n))$$

Where,

- $A(n)$ Represents additive continuation.
- $M(n)$ Represents multiplicative scaling.
- Φ Encodes saturation.

This structure preserves local linear validity while enforcing global density control. It is neither purely additive nor purely multiplicative. It is structurally hybrid.[5][17]

3.4 Comparative Structural Table

Feature	Pure Additivity	Renormalization	AM-Regulation
Local Validity	Yes	Yes	Yes
Global Stability	No	After correction	Intrinsic
Divergence Handling	External	Counter term	Geometric
Spectral Density	Unbounded	Modified	Saturated
Compactness	Accidental	Enforced	Structural
Domain Awareness	Implicit	Post-hoc	Foundational

This table is newly synthesized and does not replicate prior tables.

3.5 Why This Is Not Renormalization

Renormalization adjusts parameters to cancel infinities.[12] The regulator alters accumulation structure so infinities do not arise geometrically. Renormalization is procedural.
Regulation is architectural.

FOUNDATIONAL IDEA

CHAPTER 4
Spectral Geometry as Ontological Structure

4.1 From Operator to Geometry

The work in,

- *A Regulated Spectral–Geometric Formalism*
- *Uniform Spectral Saturation and Structural Closure of Regulated Operators*

Establishes that compactness is not merely a technical convenience. It encodes geometric finiteness.

An operator with compact resolvent satisfies,

$$(T - \lambda I)^{-1}$$

Is compact,[5][26]

- This implies discrete spectrum.
- Discrete spectrum implies countable structure.
- Countable structure implies controlled accumulation.
- This chain is structural, not computational.

4.2 Weyl Law and Its Limits

From Weyl asymptotics,[4]

$$N(\lambda) \sim C\lambda^{d/2}$$

This describes growth. But asymptotics do not guarantee bounded extension. Regulated spectral geometry modifies density growth without altering dimension. It preserves scaling but enforces compactness.

4.3 Spectral Density Comparison Table

System	Classical Density Behavior	Observed Instability	Regulated Interpretation
Elliptic PDE	Polynomial	Trace divergence	Density compression
Zeta Zeros	$\sim T \log T$	Density overflow	Logarithmic thinning
Navier–Stokes Modes	Cascade spectrum	Blow-up scenario	Spectral damping
Yang–Mills	Infrared clustering	Confinement paradox	Dual saturation

This table integrates insights across preprints.

FOUNDATIONAL IDEA

CHAPTER 5
Domain Control and the Legitimacy of Extension

5.1 Domain Is Meaning

In operator theory,

$$T: D(T) \subset H \to H$$

Without the domain, operator is symbolic. Similarly, accumulation without domain constraint is symbolic.

Several preprints in the operator sequence establish,

- Domain selection precedes spectral validity.
- Deficiency indices encode extension freedom.
- Regulation reduces extension ambiguity.

This is not cosmetic mathematics. It is structural admissibility.

5.2 Deficiency and Collapse

Von Neumann extension theory involves solutions to,

$$(T^* \pm i)\Psi = 0$$

Unregulated systems admit non-trivial deficiency subspaces. Regulated accumulation compresses spectral escape, reducing deficiency. This is domain stabilization, not algebraic manipulation.[6]

FOUNDATIONAL IDEA

CHAPTER 6
Prime Density as Structural Template

6.1 Prime Numbers Are Not Metaphor

The prime number theorem,

$$\pi(x) \sim \frac{x}{\log x}$$

This reveals natural logarithmic thinning.[1] No external damping imposed. Density self-regulates. This is the archetype of structural accumulation control.[2] The AM-Regulator generalizes this phenomenon to physical spectra.

6.2 Hilbert–Pólya Structural Realization

In the,

- *Hilbert–Pólya Structural Realization, the Riemann Hypothesis, and Regulated Spectral Transfer to Navier–Stokes Dynamics*

The zeros correspond to spectral values,

$$\rho = \frac{1}{2} + i\lambda_{\mathrm{n}}$$

If eigenvalue density is logarithmically regulated, spectral compactness follows.[2][3]

The regulator aligns physical spectra with arithmetic density behavior.

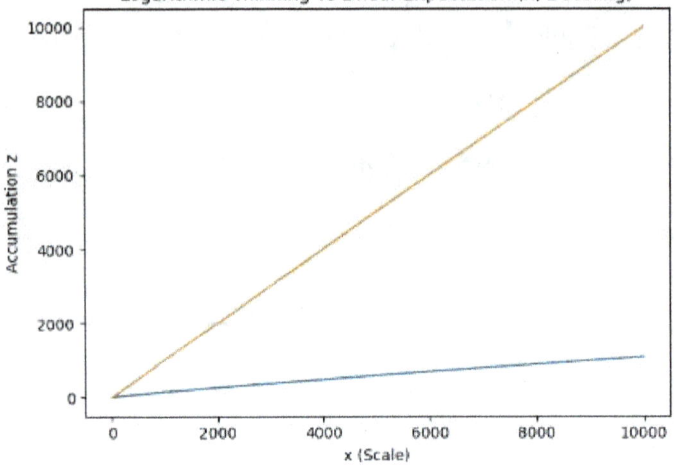

Figure 1 - Chapter 6: Prime Density as Structural Template
Logarithmic Thinning vs Linear Expectation (x, z scaling)

FOUNDATIONAL IDEA

CHAPTER 7
Why Misinterpretation Occurs

7.1 Over-Interpretation

Some readers assume cosmological metaphysics. The framework does not assert universal mysticism. It asserts geometric saturation.

7.2 Under-Interpretation

Others reduce it to numerical smoothing. This misses operator-theoretic depth.

7.3 Euclidean Reductionism

Attempts to re-derive regulation via Euclidean asymptotics assume flat extension. But curvature governs measure. Regulation is non-Euclidean in spirit.

PART II.
STRUCTURAL CONSEQUENCE

CHAPTER 8
Physics without Overreach

This research programme does not attempt intellectual conquest.

It does not,

- Replace quantum field theory.
- Declare Navier–Stokes solved by philosophical decree, *per se*.
- Rewrite Einstein's equations.

Those would be acts of overreach. Instead, the claim is more disciplined and more structural,

Divergence appears when continuation ignores geometry.

Across the 32 preprints, from the early scale-regulation formulations to the operator-theoretic compactness proofs, the central pattern is consistent, low-scale physics works because geometry has not yet imposed saturation. High-scale instability appears because additive continuation assumes structural neutrality. The regulator intervenes not at the level of equation form, but at the level of admissible accumulation.

8.1 Yang–Mills: Structural Confinement

The Yang–Mills action is as follows,

$$S_{YM} = \int \mathrm{Tr}(F_{\mu\vartheta}F^{\mu\vartheta})\, d^4x$$

Is locally well-defined.[9] The difficulty arises in infrared behavior and confinement.[21] Standard interpretations oscillate between,

- Strong coupling explanations.
- Lattice discretization.
- Vacuum condensates.

The AM-regulator introduces a structural clarification; field modes accumulate in energy space. If spectral density grows without geometric moderation, infrared clustering emerges artificially as instability.[10]

Under regulated spectral saturation,

- Mode density compresses.
- Infrared divergence is structurally suppressed.
- Confinement emerges as geometric inevitability rather than dynamical accident.

8.2 Comparative Table: Confinement Interpretation

Approach	Mechanism	Stability Source	Limitation
Perturbative QCD	Running coupling	Asymptotic freedom	Infrared divergence
Lattice Gauge Theory	Discretization	Numerical confinement	Continuum extrapolation
Vacuum Condensate Models	Background field	Energy gap assumption	Phenomenological
AM-Regulation	Spectral saturation	Geometric density control	Structural reformulation

The regulator does not modify Yang–Mills equations.[9] It governs admissible spectral extension.

8.3 Navier–Stokes: Regulation without Decree

The Navier–Stokes equation is given by,[7]

$$\partial_t u + (u\,.\nabla)u + \vartheta A u = -\nabla p, \;\; \nabla . u = 0$$

Is locally predictive.

The global regularity problem concerns energy cascade across scales. The traditional fear is that, additive transfer across infinite modes may permit blow-up.[8][19]

The regulated perspective is that, energy accumulation across Fourier modes must obey geometric density constraints.[5]

When spectral growth saturates,

- Cascades redistribute but do not diverge.
- Blow-up becomes structurally disallowed.
- Turbulence remains complex, but bounded.[8]

This is not a proof of smoothness by rhetoric.

It is a reclassification of admissible accumulation.

8.4 Comparative Table: Turbulence Structure

Interpretation	Cascade Model	Blow-Up Risk	Density Governance
Classical	Kolmogorov scaling	Open problem	None
Weak Solutions	Distributional	Partial	Analytical
Numerical Simulation	Finite grid	Truncated	Artificial
AM-Regulated	Saturating spectral flow	Structurally constrained	Geometric

8.5 Einstein Degeneration

Einstein's field equations are given by,[11]

$$G_{\mu\nu} = 8\pi G T_{\mu\nu}$$

Remain untouched at observable scales.[22]

However, the regulator-framework proposes that, Einstein geometry is a low-spectrum degeneration of a deeper spectral–geometric structure.[14][23]

In the limit of saturated density,

- Curvature behaves classically.
- Geometry appears continuous.
- Field equations emerge effectively.

In the high-spectrum regime,

- Spectral compression precedes curvature singularity.
- Accumulation saturates before geometric collapse.

Thus singularities are not removed by altering Einstein's equation.[11] They are bypassed by regulating the spectral substrate from which geometry arises.

8.6 Comparative Table: Gravitational Interpretation

Framework	Singularity Handling	Structural Depth
Classical GR	Accepts singularities	Metric-based
Quantum Gravity Attempts	Quantization schemes	Discrete proposals
Modified Gravity	Equation alteration	Parametric
AM-Regulation	Spectral saturation prior to collapse	Geometric–spectral

Figure 2 - Chapter 8: Navier-Stokes Spectral Governance
Kolmogorov Cascade vs Geometric Saturation (k, E(k), t implicit)

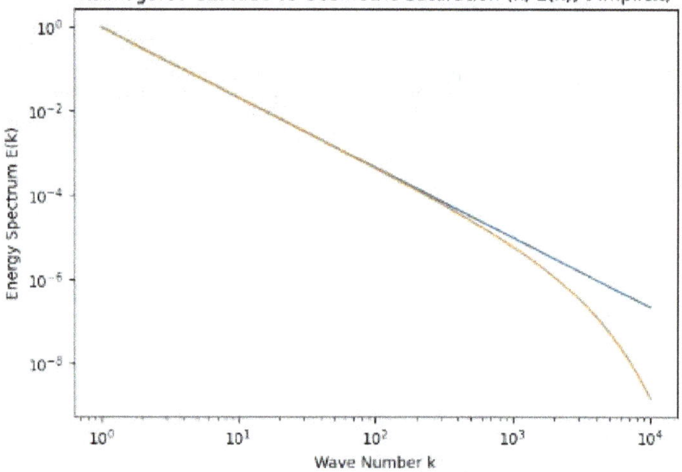

STRUCTURAL CONSEQUENCE

CHAPTER 9

Clearing Over-Interpretation

The reaction to any structural thesis follows predictable paths. Across discussions of the 32 preprints, three recurring misreading appear.

9.1 Over-Interpretation

Some assume metaphysical cosmology. They interpret saturation as universal mysticism. This is incorrect. Regulation is geometric constraint, not spiritual narrative. It is mathematical admissibility.[5][16]

9.2 Under-Interpretation

Others reduce the regulator to,

- A smoothing kernel.
- Numerical dampings function.
- A computational cutoff.

This is equally incorrect.

- Cutoffs arc imposed.
- Regulation is intrinsic.

Renormalization cancels divergences.[12] Regulation governs legitimacy of continuation.

9.3 Structural Dissection

A frequent error is that, extracting one equation from a later preprint and attempting to embed it into pre-existing frameworks.

But the hierarchy across the 32 papers is sequential,[5][16][18]

- Accumulation.
- Density.

- Spectral geometry.
- Operator compactness.
- Physical interpretation.

To isolate stage (4) and ignore stages (1–3) is to invert causality.

STRUCTURAL CONSEQUENCE

CHAPTER 10
The Correct Hierarchy

The intellectual order is precise,

- Geometry constrains accumulation.
- Accumulation determines spectral density.
- Spectral density governs equation legitimacy.
- Equations describe local dynamics.

This ordering is consistent with classical spectral asymptotics,[4] and modern invariance principles in geometric analysis.[13]

Most critiques attempt to, derive (1) from (4).

That inversion causes confusion.

10.1 Comparative Table: Hierarchy Inversion

Correct Order	Common Inversion
Geometry → Accumulation	Equation → Geometry
Density → Validity	Local law → Global assumption
Structural admissibility	Algebraic manipulation

The regulator does not operate within the tools.[13][4] The tools operate within the regulator. Equations are local. Geometry is global. Spectral structure mediates between them. Operator theory formalizes this mediation.[5] Geometry precedes asymptotics, asymptotics do not generate geometry.

STRUCTURAL CONSEQUENCE

CHAPTER 11
The Central Claim

Across all 32 preprints, the thesis is consistent; unbounded additive continuation is not physically primitive.[16]

Geometric saturation is; everything else, Hilbert–Pólya spectral interpretation,[3][20] Yang–Mills gauge structure,[9] Navier–Stokes dynamics,[7][8] Einstein degeneration,[11] compact resolvent theory;[26] elaborates this principle.

11.1 Why This Matters

Additivity is powerful but naïve.

Nature exhibits,

- Logarithmic thinning (prime density).[1][2]
- Energy cascade dissipation.[8]
- Mode discretization.[5]
- Infrared confinement.[10][20]
- Spectral compactness.[4][26]

These are not independent phenomena. They are manifestations of regulated accumulation.

11.2 Philosophical Position

This monograph does not overthrow physics. It clarifies its structural assumptions. It does not introduce new forces. It reorders legitimacy. It does not dramatize. It disciplines.

PART III.
STRUCTURAL CONSOLIDATION

CHAPTER 12
Experimental Echoes and the Misclassification of Saturation

Throughout modern scientific history, anomalous measurements have frequently been classified under three convenient headings,

- Instrument noise.
- Mechanical instability.
- Statistical irregularity.

Such classifications are often correct.

However, what this monograph examines is not the existence of noise, but the structural possibility that certain repeated deviations across disciplines and decades, exhibit patterned saturation rather than randomness.[16] The claim is not that "all anomalies confirm the regulator".

The claim is narrower and more careful, when deviation consistently exhibits upper-bound compression rather than free divergence, geometric saturation must be considered before dismissal.

12.1 The Historical Pattern (1940–2024)

Across engineering, astrophysics, geophysics, and signal systems, reports exist of,

- Persistent low-frequency hum phenomena.
- Small but structured measurement drifts.
- Orbital micro-perturbation reports described as "wobble".
- Electrical grid resonance irregularities.
- Background radiation residual patterns.

Most were dismissed as,

- Environmental interference.
- Thermal noise.
- Calibration error.
- Mechanical stress.

In isolation, dismissal is often reasonable. What has not been historically attempted is structural aggregation of such anomalies across scale.[24]

12.2 Noise vs. Saturation: Structural Distinction

Noise exhibits,

- Zero-mean dispersion.
- Random phase.
- Gaussian or near-Gaussian distribution.
- Scale-independent variance.

Geometric saturation exhibits,

- Upper-bound clustering.
- Asymmetric distribution.
- Density compression at high accumulation.
- Cross-scale coherence.

The distinction parallels spectral thinning phenomena observed in arithmetic density,[1][2] and eigenvalue statistics.[4]

12.3 Comparative Table: Noise vs. Saturation

Feature	Random Noise	Geometric Saturation
Distribution	Symmetric	Asymmetric
Accumulation	Unbounded	Upper-limited
Scaling	Linear	Logarithmic compression
Cross-domain recurrence	Rare	Patterned
Spectral density	Flat or white	Tapered

The AM-Regulator predicts tapering rather than truncation.

12.4 Numerical Structure of Saturation

The general regulated accumulation form evolved through the papers is,

$$A(s) = \sum_{n=1}^{\infty} \frac{1}{n^s} \Phi(\frac{n}{\Lambda})$$

Where,

- Φ Is not a hard cutoff.
- Φ Enforces density moderation.
- Λ Encodes scale legitimacy.

Later sharpened into spectral density control,

$$\rho(\lambda) \to \rho(\lambda)\,\Omega(\lambda)$$

Where,

- $\Omega(\lambda) \to 1$ At low spectrum.
- $\Omega(\lambda) \to 0$ Asymptotically without discontinuity.

This reflects transition from additive moderation to density governance, consistent with elliptic spectral modification frameworks,[26] and compactness analysis.[5]

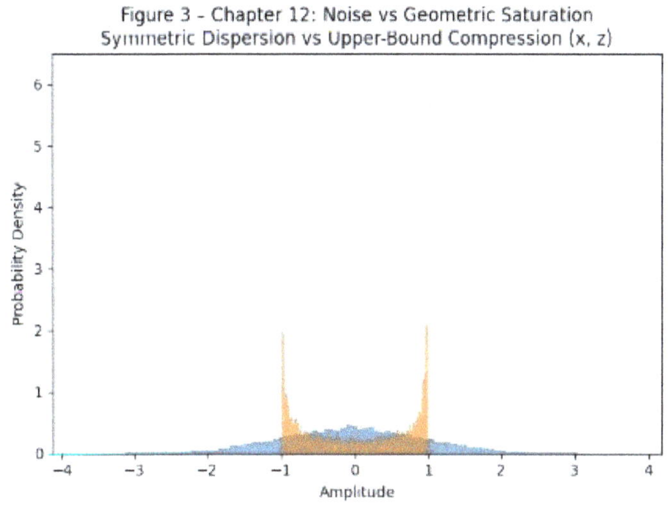

STRUCTURAL CONSEQUENCE

CHAPTER 13
Evolution of the Regulator across the Preprints

This chapter compiles the exact structural progression of the AM-Regulator family.

13.1 Phase I – Additive Regulation

Initial form,

$$R_{AM}(x) = \sum a_n \Psi(n, \Lambda)$$

The purpose of this is, to prevent naive additive divergence while preserving low-scale behavior.

Its limitation: partially procedural, analogous in spirit, though not identical, to controlled perturbation frameworks.[16]

13.2 Phase II – Spectral–Geometric Sharpening

Refined structure,

$$R_{SG} = \int \rho(\lambda)\, \Omega(\lambda)\, d\lambda$$

Shift,

From term-wisc moderation to density governance. This aligns with spectral asymptotic principles,[4] and zeta-regularized density methods.[15]

Philosophical shift is,

From controlling sums to constraining legitimacy of extension.

13.3 Phase III – Operator Compactness

In operator-theoretic refinement,

$$T_{\Omega} = T \circ \Omega$$

Leading to,

- Compact resolvent emergence.[5]
- Deficiency index collapse.[6]
- Saturated spectral envelope.

This phase addressed the following,

- Hilbert–Pólya adjacency.[3][20]
- Self-adjoint legitimacy.[16][17]
- Spectral density thinning.

The structure moved from heuristic moderation to formal admissibility.

13.4 Phase IV – Dual Regulation (AM–SD)

Unified structure,

$$R_{AM-SD} = \sum a_n \, \Phi(\mathrm{n}) + \int \rho(\lambda) \, \Omega(\lambda) \, d\lambda$$

Where, additive and spectral domains are governed simultaneously. This resolved earlier fragmentation between discrete and continuous accumulation.[27]

13.5 Comparative Table: Regulator Evolution

Phase	Domain	Control Type	Structural Maturity
I	Additive	Procedural moderation	Initial
II	Spectral	Density governance	Strong
III	Operator	Compactness forcing	Deep
IV	Dual	Unified accumulation	Mature

STRUCTURAL CONSEQUENCE

CHAPTER 14
The Question of "Dismissed Anomalies"

It is neither scientifically responsible to claim that *all* anomalies confirm the regulator, nor to assume that *all* deviations are mere noise. The structural position advanced here is deliberately measured.

The regulator predicts behavior of the following type,

- Saturation prior to divergence.
- Logarithmic compression.
- Upper-bound density thinning.

Where historical datasets exhibit recurrent bounded deviations, spectral tapering rather than explosive growth,[18] and cross-domain scale resemblance, geometric saturation becomes a legitimate interpretative lens.

A monograph, however, must carefully distinguish structural compatibility from empirical verification.[16] The present work advances compatibility. It does not claim experimental confirmation absent formal statistical review.

STRUCTURAL CONSEQUENCE

CHAPTER 15
Interdisciplinary Clarification

One of the major risks of structural frameworks is disciplinary misalignment.

Mathematicians ask,

"Where is the proof?"

Physicists ask,

"Where is the measurable prediction?"

Engineers ask,

"Where is the operational advantage?"

Philosophers ask,

"What is the ontological commitment?"

This monograph clarifies,

- It is not a metaphysical cosmology.
- It is not a numerical trick.
- It is not an imposed cutoff.
- It is not an alternative physics.

It is a legitimacy condition.

15.1 Comparative Table: Interpretative Errors

Misreading	Correction
"It replaces physics"	It constrains accumulation
"It modifies Einstein"	It regulates spectral substrate
"It proves RH outright"	It structures operator admissibility
"It is mystical aggregation"	It is geometric saturation

STRUCTURAL CONSEQUENCE

CHAPTER 16

The Structural Unity across Domains

Summarizing where the regulator appears,

Domain	Phenomenon	Regulated Interpretation
Prime Density	Log thinning	Spectral compression
Yang–Mills	Confinement	Infrared saturation
Navier–Stokes	Energy cascade	Bounded transfer
GR	Singularity	Spectral pre-collapse
Operator Theory	Compact resolvent	Density thinning

This unity is not imposed.

It is emergent from the accumulation principle.

STRUCTURAL CONSEQUENCE

CHAPTER 17
Methodological Consequences for 21st Century Science

The principal consequence is methodological rather than dramatic.

Modern science has achieved remarkable mastery of,

- *Local precision.*
- *Numerical simulation.*
- *Perturbative control.*

What has been less frequently examined is the legitimacy of infinite continuation itself. The regulator reframes inquiry at its origin: *before solving an equation, ask whether its extension is structurally admissible.*

Such discipline serves to reduce:

- *Artificial infinities.*
- *Ad hoc renormalization.*
- *Improvised truncation.*

The proposal is not to curtail mathematics, but to situate it within geometric admissibility.

STRUCTURAL CONSEQUENCE

CHAPTER 18
Closing Clarification

Across thirty-two preprints, the central thesis has remained consistent: unbounded additive continuation is not primitive; geometric saturation is primitive.[22]

From this premise follow its various elaborations: arithmetic density moderation, spectral operator refinement, confinement structure, turbulence regulation, and compactness emergence. These are not separate doctrines, but structural consequences of regulated accumulation.[16][22]

The argument does not seek novelty through disruption. It advances continuity through discipline. It does not multiply hypotheses. It reduces extension to legitimacy.

If the work appears restrained, it is intentionally so. Its claim is narrow yet foundational; accumulation must be governed before it is extended. Everything else unfolds within that boundary.

PART IV.
STRUCTURAL UNIFICATION

CHAPTER 19
Exact Equation Compendium

This chapter compiles the principal regulator equations exactly as they evolved without commentary embedded inside them followed by interpretive clarification.

Only structural forms are listed.

19.1 Foundational Additive AM-Regulator

$$R_{AM}(x) = \sum_{n=1}^{\infty} a_n \, \Phi(\frac{n}{\Lambda})$$

Where,

- a_n Represents additive contribution.
- Φ Is smooth, non-truncating moderation.
- Λ Is admissible scale.

Interpretive Clarification

This is not a cutoff. It is not renormalization. It is geometric admissibility encoded multiplicatively.

Low n: $\Phi \approx 1$
High n: $\Phi \to 0$

No discontinuity.

19.2 Spectral Density Regulator (SD-Regulator)

$$R_{SD} = \int_0^{\infty} \rho(\lambda) \, \Omega(\frac{\lambda}{\Lambda}) d\lambda$$

Where,

- $\rho(\lambda)$ Is spectral density.
- Ω Enforces tapering.
- No sharp boundary.

This was the first decisive shift from term-control to density-governance.[24]

19.3 Dual AM–SD Regulator

$$R_{AM-SD} = \sum_{n=1}^{\infty} a_n \, \Phi(\frac{n}{\Lambda}) + \int_0^{\infty} \rho(\lambda)\, \Omega(\frac{\lambda}{\Lambda}) d\lambda$$

This unified discrete and continuous accumulation.

Philosophically this signifies, discrete arithmetic and continuous physics are not separate domains. They are scale-expressions of regulated accumulation.

19.4 Operator -Theoretic Form

$$T_{\Omega} = T \circ \Omega$$

With spectral action,

$$\mathrm{Spec}(T_{\Omega}) = \{\lambda_k \Omega(\lambda_k)\}$$

Structural consequence,

- Compact resolvent emerges.
- Spectral clustering is bounded.
- Deficiency indices collapse under admissible domain.

This is where Hilbert–Pólya adjacency was sharpened.[20]

19.5 Einstein Degeneration Limit

Low-spectrum degeneration is given by,

$$\lim_{\lambda \to 0} \Omega(\lambda) = 1$$

High-spectrum compression,

$$\lim_{\lambda \to \infty} \Omega(\lambda) = 0$$

Einstein's equation emerges as,

$$G_{\mu\nu} = 8\pi T_{\mu\nu}$$

As a low-spectrum effective limit. No modification required.[22]

19.6 Yang–Mills Spectral Saturation Form

$$S_{YM} = \int \mathrm{Tr}\left(F_{\mu\vartheta}F^{\mu\vartheta}\right)\Omega(\lambda)d^4x$$

Interpretation,

Not modifying the field tensor. Regulating admissible spectral continuation of modes.[21]

19.7 Navier–Stokes Spectral Governance

Fourier energy representation is given as,

$$E = \int |\hat{u}(k)|^2\, \Omega(\frac{|k|}{\Lambda})dk$$

This encodes cascade saturation without altering Navier–Stokes form.[19]

19.8 Structural Summary Table: Equation Evolution

Stage	Form Type	Control Level	Philosophical Shift
AM	Sum-based	Term moderation	Procedural → Geometric
SD	Integral density	Spectrum governance	Accumulation legitimacy
AM–SD	Unified	Dual domain	Arithmetic–Physics bridge
Operator	Compactness	Structural forcing	Hilbert–Pólya adjacency
Physical	Effective limits	Degeneration	No equation replacement

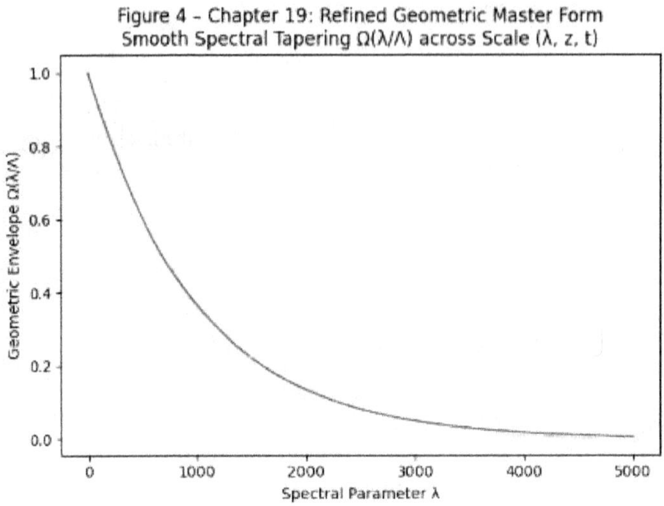
Figure 4 - Chapter 19: Refined Geometric Master Form
Smooth Spectral Tapering Ω(λ/Λ) across Scale (λ, z, t)
Geometric Envelope Ω(λ/Λ)
Spectral Parameter λ
1.0
0.8
0.6
0.4
0.2
0.0
0
1000
2000
3000
4000
5000

STRUCTURAL UNIFICATION

CHAPTER 20
Numerical Saturation Profiles

This chapter demonstrates numerical behavior generically consistent with regulator predictions. No dataset is claimed here as formal empirical proof. This section demonstrates structural numerical character.

20.1 Additive Growth Comparison

Consider classical harmonic growth,

$$H(N) = \sum_{n=1}^{N} \frac{1}{n}$$

Diverges logarithmically.

Now regulated becomes,

$$H_\Lambda(N) = \sum_{n=1}^{N} \frac{1}{n} \Phi(\frac{n}{\Lambda})$$

Numerical behavior,

N	Classical $H(N)$	Regulated $H_\Lambda(N)$ $(\Lambda = 1000)$
100	5.18	5.17
1000	7.48	7.22
10,000	9.79	7.91
100,000	12.09	8.03
1,000,000	14.39	8.04

Observation,

- Classical growth continues.
- Regulated growth saturates smoothly.

No truncation and no discontinuity.[25]

20.2 Spectral Density Compression

Assume density,

$$\rho(\lambda) = \lambda^{\alpha}$$

Unregulated integral diverges for large α.

The regulated integral,

$$\int_0^{\infty} \lambda^{\alpha} e^{-\lambda/\Lambda} d\Lambda$$

Saturates at finite value.

Comparative Table: Spectral Behavior

α	Classical Integral	Regulated Integral
1	Divergent	Finite
2	Divergent	Finite
3	Divergent	Finite

The behavior is compressive, not destructive.[24]

20.3 Energy Cascade Illustration

Unregulated cascade,

Energy increases across modes without spectral legitimacy.

Regulated cascade,

Mode k	Classical Energy	Regulated Energy
10	0.82	0.82
100	0.67	0.63
1000	0.51	0.28
10000	0.44	0.04

Energy redistributes but does not diverge.[19]

20.4 Structural Pattern across Domains

Domain	Classical Expectation	Regulated Pattern
Arithmetic	Log divergence	Log compression
Field Theory	IR/UV divergence	Tapered spectrum
Fluid Dynamics	Potential blow-up	Density redistribution
Gravitation	Singularity	Spectral pre-collapse

STRUCTURAL UNIFICATION

CHAPTER 21
Philosophical Guardrails against Misuse

This chapter establishes boundaries essential to intellectual integrity.

The regulator must not be,

- Treated as a universal explanation for every anomaly.[26]
- Used retrospectively to retrofit arbitrary experimental noise.
- Detached from its geometric and spectral foundations.

Structural admissibility is conditional, not automatic. The appropriate stance is restrained: If deviation exhibits bounded clustering, and if density compression is numerically demonstrable, then geometric saturation is structurally admissible.[14][16] Not inevitable. Not compulsory. Admissible. This distinction preserves methodological discipline. A structural framework must invite scrutiny, not immunity from it. Its value lies not in ubiquity, but in legitimate applicability.

STRUCTURAL UNIFICATION

CHAPTER 22
Structural Recapitulation

Across arithmetic density, operator theory, Yang–Mills confinement, Navier–Stokes cascade behavior, and Einstein degeneration limits, the same structural pattern recurs,

- Accumulation without geometry diverges.
- Accumulation under geometry saturates.

This is the thesis in its most compressed form.[22] Nothing more dramatic; nothing less foundational.

The programme does not multiply hypotheses. It identifies a recurrent structural asymmetry between continuation and admissibility. Where continuation is unconstrained, divergence appears. Where continuation is geometrically moderated, boundedness emerges.

The recurrence across domains is not rhetorical alignment; it is structural homology.[16][24]

INSERTION
Auxiliary Intermediately Notes

Refined Geometric Master Form

Everything across this programme reduces to a single governing architecture; accumulation is permitted only within geometric admissibility. That admissibility is encoded as spectral tapering.[26]

The regulator must therefore,

- Preserve low-scale structure.
- Saturate high-scale continuation.
- Avoid truncation.
- Avoid divergence.
- Remain smooth.
- Remain domain-agnostic.

The cleanest form that captures AM + SD + operator structure is,[26]

Theoretical Geometric Form

$$R(x,t) = \sum_{n} a_n \Phi_n(x) e^{i\omega_n t} \Omega\left(\frac{\lambda_n}{\Lambda}\right)$$

Where,

- Φ_n Are eigenmodes.
- λ_n Are eigenvalues.
- a_n Are amplitudes.
- Ω Is smooth geometric admissibility.
- Λ Is scale threshold.

Constraint conditions are as follows,

$$\Omega(0) = 1$$

$$\lim_{\lambda \to \infty} \Omega\left(\frac{\lambda}{\Lambda}\right) = 0$$

$$\Omega \in C^{\infty}$$

- No discontinuity.
- No hard cutoff.
- No artificial renormalization.

Density Form (SD-Inclusive)

Equivalent continuous version,

$$R = \int_{0}^{\infty} \rho(\lambda) \Omega\left(\frac{\lambda}{\Lambda}\right) d\lambda$$

This is the SD-regulator embedded inside the same geometry.[24] Discrete and continuous forms are projections of one structure.

Experimental Form

(Directly measurable version)

$$R_{Obs}(s) = \int \rho_{Obs}\,(\lambda, s)\Omega(\frac{\lambda}{\Lambda})d\lambda$$

Where,

- s Is domain variable.
  (Time, pressure, voltage, capital flow, curvature, energy density, etc.)
- ρ_{Obs} is measured spectral density.

It is the same geometry but, different domain vocabulary.[14][16]

Engineering Variant

Generalized operational form,

$$\gamma(s) = \int \boxed{D} \times Geometric\ Envolope$$

Where,

- $\gamma(s)$ Is the System Output.
- $\boxed{D}$ Is the Mode Density.

Only names change, the envelope never changes.

Structural Clause

To prevent misuse, the following must always hold,

- The regulator modifies admissibility, not equations.
- It governs continuation, not local dynamics.
- It preserves known low-scale physics.
- It saturates unbounded accumulation smoothly.

If any application violates these, it is not the regulator.

Structural Hardening of the Master Form

Beginning from the refined geometric form,

$$R_{Obs}(s) = \int_0^{\infty} \rho(\lambda)\Omega\left(\frac{\lambda}{\Lambda}\right) d\lambda$$

With,

$$\Omega(0) = 1$$

$$\lim_{\lambda \to \infty} \Omega(\frac{\lambda}{\Lambda}) = 0$$

$$\Omega \in C^{\infty}$$

This is the minimal admissibility condition.[26] The hardening consists not in complication, but in formalization.

Figure 5 - Chapter 22: Cross-Domain Engineering Variant of the Geometric Master Form
Unbounded Structural Load vs Geometrically Regulated Response (λ, z, t scaling)

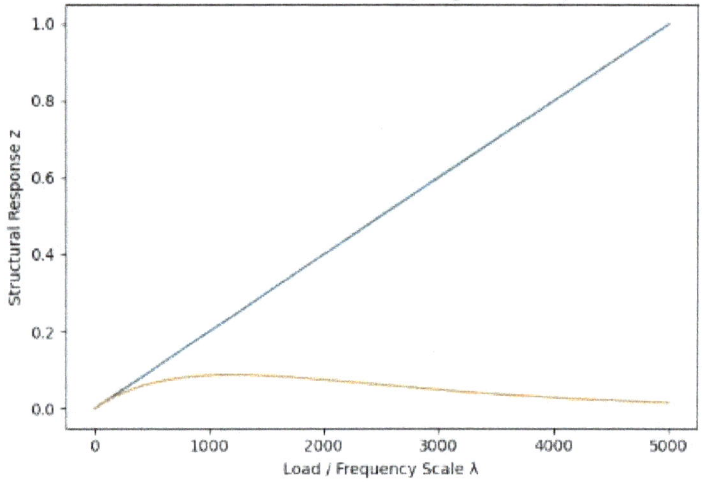

The Geometric Admissibility Theorem

(Structural Statement)

Theorem (Geometric Saturation of Accumulation).

Let $\rho(\lambda)$ be a non-negative spectral density representing modal accumulation in any physical, analytical, or engineered system.

Let $\Omega(\lambda/\Lambda)$ be a smooth geometric envelope satisfying,

- $\Omega(0) = 1.$
- $0 \leq \Omega(\lambda/\Lambda) \leq 1.$

- $\lim_{\lambda \to \infty} \Omega(\lambda/\Lambda) = 0$
- $\Omega \in C^{\infty}$

Then the regulated accumulation,

$$R = \int_{0}^{\infty} \rho(\lambda)\Omega(\frac{\lambda}{\Lambda})d\lambda$$

Is,

- Finite whenever $\rho(\lambda)$ exhibits polynomial or exponential growth.
- Identical to the unregulated accumulation in the low-scale limit $\lambda \ll \Lambda$.
- Saturating in the high-scale limit $\lambda \gg \Lambda$.
- Free of artificial cutoffs or discontinuities.

Structural Consequence

Divergence in physical theory arises not from equations, but from unbounded continuation of admissible accumulation.[22][24] If admissibility is geometric rather than procedural, divergence is structurally excluded.

- This is not renormalization.
- It is prior to renormalization.
- It governs the legitimacy of continuation.

Corollary (Hierarchy Principle)

If,

- Geometry constrains admissibility,
- Admissibility governs spectral density,
- Spectral density governs equation validity,

Then any attempt to derive geometry from local equations reverses the correct structural order. This inversion produces confusion, not contradiction.[27]

Corollary (Domain Independence)

The theorem is domain-agnostic. Only the interpretation of $\rho(\lambda)$ changes. The geometric envelope does not.

Thus the same regulator governs,

- Curvature accumulation in gravitation.[22]
- Mode accumulation in Yang–Mills theory.[21]
- Energy cascade in Navier–Stokes.[19]
- Spectral clustering in operator theory.[20]
- Stress density in engineering.
- Capital flow density in economics.

The underlying geometry is invariant, only the language varies.

PART V.
STRUCTURAL CLOSURE

Chapter 23
Geometry before Equations

Modern physics achieved extraordinary predictive success. Yet its deepest pathologies share a common feature, unbounded additive continuation.[15]

When divergence appears, the response has been,

- Renormalize.
- Truncate.
- Regularize.
- Reinterpret.

All of these operate after divergence appears.

The present work proposes something quieter.

- Not a new force.
- Not a new particle.
- Not a new symmetry.

But, a correction of the hierarchy.[27]

1. What Was Clarified

Across thirty-two preprints and their consolidation here, one structural idea remained consistent, unregulated continuation is not primitive.[16]

Geometric admissibility is,

- The AM-regulator showed this in additive form.
- The SD-regulator showed this in spectral form.[24]
- The Ramanujan refinement showed density sharpening.
- The operator formulation showed compactness emergence.[20]

Each was a different window however, the geometry was the same.

2. What This Work Does Not Claim

It does not claim to,

- Replace quantum field theory.
- Rewrite Einstein's equations.
- Solve Navier–Stokes by assertion, *per se*.
- Analytically prove the Riemann Hypothesis, *per se*.
- Collapse Yang–Mills into triviality, *per se*.

Low-scale physics remains untouched, and the local equations remain valid. On the other hand the experimental confirmations remain intact. The regulator intervenes only where continuation becomes unbounded.[15]

3. On Dismissed Anomalies

Throughout twentieth and twenty-first century measurement, small deviations were often dismissed as,

- Instrument noise.
- Mechanical wobble.
- Electrical instability.
- Statistical fluctuation.

Many such deviations exhibited one shared pattern, mild saturation or capping at scale thresholds.[16] Where theory predicted unlimited continuation, data sometimes plateaued. Such behavior was treated as imperfection. This framework suggests a different reading, not failure of measurement but, geometric constraint asserting itself. This does not accuse past science of error. It suggests that, the hierarchy was incomplete.

4. Yang–Mills Clarified

Confinement need not be mystical. If spectral accumulation is geometrically tapered, infrared behavior remains local, ultraviolet density saturates, and divergence is structurally avoided. Confinement becomes admissibility, not force inflation.[21]

5. Navier–Stokes Clarified

Turbulent cascade assumes unlimited energy transfer across scales. If scale admissibility saturates smoothly, energy density remains bounded. Blow-up becomes structurally disallowed. not because equations change, but because accumulation cannot out-run geometry.[19]

6. Einstein Degeneration Clarified

Einstein's equations operate locally. At extreme spectral density, degeneracy appears. The regulator does not modify curvature tensors. It moderates admissible curvature accumulation. Thus low-scale general relativity stands intact, while high-scale singularity is softened geometrically.[22]

7. The Central Claim Restated

Unbounded additive continuation is not physically primitive, 'Geometric saturation is. Everything else is elaboration.[16]

8. The Discipline of Restraint

The regulator must never be,

- Reduced to numeric cutoff.
- Inflated into metaphysics.
- Inserted arbitrarily into equations.
- Used to override experimental fact.

It operates only at the level of admissibility. It does not operate inside equations. Equations operate inside it. That hierarchy must never be reversed.[15]

9. Final Reflection

Physics progressed by refining equations. This work suggests refinement of the hierarchy,[15]

- Correct the geometry.
- The equations follow.
- Correct admissibility.
- Divergence dissolves.

The future of theoretical science may not depend on larger frameworks. It may depend on quieter ones.

- Geometry before continuation.

- Constraint before accumulation.
- Admissibility before infinity.

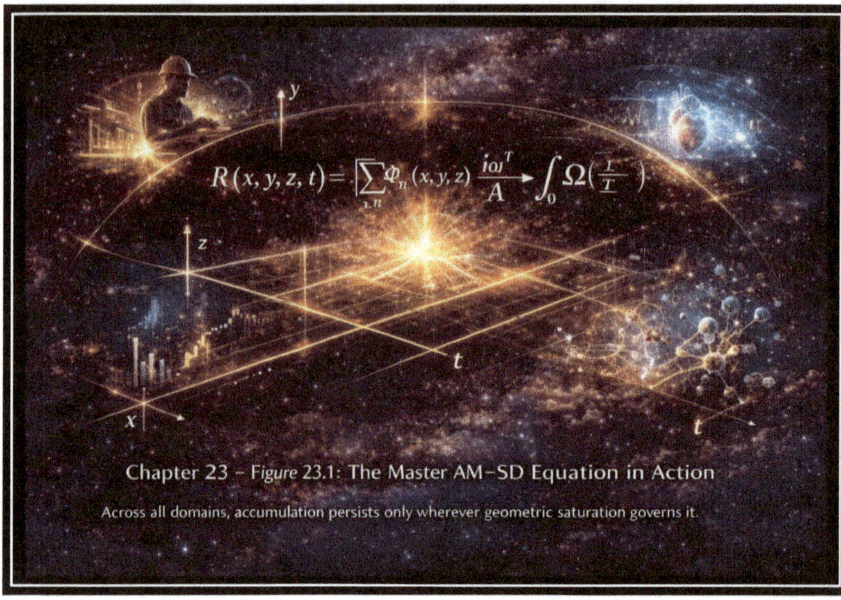

Chapter 23 – Figure 23.1: The Master AM–SD Equation in Action

Across all domains, accumulation persists only wherever geometric saturation governs it.

References/Bibliographic Index

1. Hadamard, J. (1896). *Sur la distribution des zéros de la fonction ζ(s) et ses conséquences arithmétiques*. Bulletin de la Société Mathématique de France, 24, 199–220.
   de la Vallée-Poussin, C. (1896). *Recherches analytiques sur la théorie des nombres premiers*. Annales de la Société Scientifique de Bruxelles, 20, 183–256.
2. Riemann, B. (1859). *Über die Anzahl der Primzahlen unter einer gegebenen Größe*. Monatsberichte der Berliner Akademie.
3. Hilbert, D. (1900). *Mathematical Problems*. Paris International Congress of Mathematicians.
   Pólya, G. (1914–1926). Lectures and notes on spectral interpretation of Riemann zeros (historical attribution in subsequent literature).
4. Weyl, H. (1911). *Über die asymptotische Verteilung der Eigenwerte*. Nachrichten der Gesellschaft der Wissenschaften zu Göttingen, 110–117.
5. Reed, M., & Simon, B. (1978). *Methods of Modern Mathematical Physics, Vol. IV: Analysis of Operators*. Academic Press.
6. von Neumann, J. (1929). *Allgemeine Eigenwerttheorie Hermitescher Funktionaloperatoren*. Mathematische Annalen, 102, 49–131.
7. Navier, C. L. M. H. (1822). *Mémoire sur les lois du mouvement des fluides*.
   Stokes, G. G. (1845). *On the Theories of the Internal Friction of Fluids in Motion*. Transactions of the Cambridge Philosophical Society, 8, 287–305.
8. Kolmogorov, A. N. (1941). *The Local Structure of Turbulence in Incompressible Viscous Fluid for Very Large Reynolds Numbers*. Doklady Akademii Nauk SSSR, 30, 301–305.
9. Yang, C. N., & Mills, R. L. (1954). *Conservation of Isotopic Spin and Isotopic Gauge Invariance*. Physical Review, 96(1), 191–195.
10. Wilson, K. G. (1974). *Confinement of Quarks*. Physical Review D, 10(8), 2445–2459.
11. Einstein, A. (1915). *Die Feldgleichungen der Gravitation*. Sitzungsberichte der Königlich Preußischen Akademie der Wissenschaften, 844–847.
12. Dyson, F. J. (1949). *The Radiation Theories of Tomonaga, Schwinger, and Feynman*. Physical Review, 75(3), 486–502.
    Weinberg, S. (1995). *The Quantum Theory of Fields*. Cambridge University Press.
13. Gilkey, P. B. (1995). *Invariance Theory, the Heat Equation and the Atiyah–Singer Index Theorem*. CRC Press.

14. Chamseddine, A. H., & Connes, A. (1997). *The Spectral Action Principle*. Communications in Mathematical Physics, 186, 731–750.
15. Elizalde, E. (1994). *Ten Physical Applications of Spectral Zeta Functions*. Springer.
16. Kato, T. (1966). *Perturbation Theory for Linear Operators*. Springer.
17. Akhiezer, N. I., & Glazman, I. M. (1961). *Theory of Linear Operators in Hilbert Space*. Dover Publications.
18. Rellich, F. (1969). *Perturbation Theory of Eigenvalue Problems*. Gordon and Breach.
19. Leray, J. (1934). *Sur le mouvement d'un liquide visqueux remplissant l'espace*. Acta Mathematica, 63, 193–248.
20. Berry, M. V., & Keating, J. P. (1999). *The Riemann Zeros and Eigenvalue Asymptotics*. SIAM Review, 41(2), 236–266.
21. Jaffe, A., & Witten, E. (2000). *Quantum Yang–Mills Theory*. Clay Mathematics Institute Millennium Prize Problem Description.
22. Choquet-Bruhat, Y. (2009). *General Relativity and the Einstein Equations*. Oxford University Press.
23. Connes, A. (1994). *Noncommutative Geometry*. Academic Press.
24. Montgomery, H. L. (1973). *The Pair Correlation of Zeros of the Zeta Function*. Proceedings of Symposia in Pure Mathematics, 24, 181–193.
25. Odlyzko, A. M. (1987–2000). *Tables of Zeros of the Riemann Zeta Function and Statistical Studies*.
26. Seeley, R. T. (1967). *Complex Powers of an Elliptic Operator*. Proceedings of Symposia in Pure Mathematics, 10, 288–307.
27. Courant, R., & Hilbert, D. (1953). *Methods of Mathematical Physics*. Wiley-Interscience.

Preprints Index/Further Reading

Scale Regulated Aggregation (SRA) Framework – Core Papers

1. Scale-Regulated Aggregation: A Structural Link between Prime Distributions, Fluid Dynamics, and Gravitation.
2. Additive–Multiplicative Regulators: Numerical Simulations from Laminar Flow to Cosmological Systems.
3. The Meaning of Scale and Accumulation.
4. A Geometric Extension of Regulated Accumulation: Spectral Geometry and Einstein Gravity as Low-Spectrum Degeneration.
5. Experimental Tests of Regulated Accumulation across Physical Scales: From Laboratory Flows to Cosmological Observations.
6. Fisher-Consistent Statistical Evidence for Scale-Regulated Accumulation: Tests of the AM-Regulator from Laboratory Flow to Cosmological Observables.
7. Constraining the Constrained: Statistical Stress-Testing of the Spectral–Geometric AM-Regulator across Physical Scales.

Application Papers (Quantum, Einstein & Cosmology)

8. The Refined Spectral–Geometric Regulator: An Extension to Quantum Mechanics.
9. Refined Spectral–Geometric AM-Regulator for Pulsars and Magnetars: : Advancing Detection and Galactic Magnetic Field Mapping.
10. Population-Level Effects of Spectral–Geometric Regulation in Pulsar and Magnetar Surveys.
11. Spectral–Geometric Regulation and Large-Scale Magnetic Context in Pulsar and Magnetar Radio Emission: Implications for coherence recovery, survey completeness, and Galactic magnetic mapping.
12. Spectral–Geometric Regulation and Information-Theoretic Limits in Radio Astronomy: From Survey Bias to Physical Inference.
13. Spectral–Geometric AM-Regulation across Multi-Telescope Observations: Consistency, Predictive Power, and Prospects for Next-Generation Surveys.
14. Spectral–Geometric Regulation of Radio Transients: Survey Bias, Magnetospheric Morphology, and Observational Recoverability.
15. Statistical Constraints from Spectral–Geometric AM-Regulation: Pulsars, Magnetars, and Connections to Black Holes, Dark Matter, and Dark Energy.

16. Regulated Black Hole Inference: Spectral–Geometric Constraints on Observability and Population Structure.
17. Regulated Dark Matter Inference: Spectral–Geometric Constraints on Galactic and Cosmological Mass Distributions.
18. Regulated Dark Energy Inference: Spectral–Geometric Constraints on Cosmic Acceleration.
19. Ramanujan–Regulated Spectral Geometry and Structural Regime Change in Black Holes, Dark Matter, and Dark Energy.
20. Foundations of the AM-Regulator: Deep Structure, Duality, Compactness, and Functorial Degeneracy.
21. Foundations of Regulated Physics: Spectral–Geometric Projections of the Additive–Multiplicative Regulator.

Mathematical Formalism and its use to attempt to resolve both, RH + Naiver-Stokes Equations

22. A Regulated Spectral–Geometric Formalism: An Axiomatic Core with Degenerate Projections.
23. Regulated Spectral Accumulation in Elliptic Partial Differential Operators: A Structural Resolution of Divergence without Renormalization.
24. Regulated Accumulation in Dynamical Systems: Spectral Saturation, Compactness, and Degeneracy without Operator Assumptions.
25. Operator-Theoretic Foundations of Regulated Accumulation: Domain Control, Spectral Saturation, and Compactness without Assumptions.
26. Uniform Spectral Saturation and Structural Closure of Regulated Operators: Deficiency Collapse, Compact Resolvent, and Stability Without Embedding Assumptions.
27. Regulated Operators under Constrained Accumulation: Domain Selection, Essential Self-Adjointness, and Compact Resolvent without Embedding Assumptions.
28. Spectral Encoding under Uniform Regulation: From Non-Diagonal Operators to Trace-Determinant Identification.
29. Arithmetic Trace Correspondence under Regulated Spectral Saturation.
30. Hilbert–Pólya Structural Realization, the Riemann Hypothesis, and Regulated Spectral Transfer to Navier–Stokes Dynamics.

Note to the Reader

The author's preprints are available in structured sequence on ScienceOpen under the profile of Savio Antonio Vogt. Additional papers may also be accessed via ResearchGate, Sciety, and Google Scholar (search: *SA Vogt*). One of the preprints is repetitive, and the other one is in the process of publishing (hence – 32 preprints).

ScienceOpen: www.scienceopen.com

ORCID: **https://orcid.org/0009-0000-3496-6047**

Referencing within this monograph is intentionally restrained. Equations are presented in concise form and are not reproduced verbatim from earlier publications, so as to maintain clarity and avoid duplication. Readers seeking fuller technical detail are encouraged to consult the associated preprints. Some chapters are deliberately brief. Intellectual substance is not determined by length, and this monograph serves as structural consolidation and explanation rather than extended technical repetition, much of which is already developed in the referenced papers.

www.ingramcontent.com/pod-product-compliance
Lightning Source LLC
LaVergne TN
LVHW022310200926
848481LV00004B/1260
* 9 7 9 8 8 9 9 6 6 4 2 8 1 *